POLLINATION OF FLOWERS BY BEES

Showing bees as honey and pollen gatherers and fertilizers of flowers
(See Chapter IX)

1, Violet, part of corolla cut away ; 2, horse-chestnut ; 3, orchis (the bee
is carrying away the two pollen masses or "pollinia", which will strike the
sticky stigma of the next flower it visits); 4, white dead-nettle, with stigma
touching bee's back ; 5, willow ; 6, yellow toad-flax ; 7, 8, long-styled and
short-styled forms of primula species. In 8 the flower depends on the visits
of insects for pollination.

BEE-KEEPING
FOR BEGINNERS

BY

I. H. JACKSON, M.A.

Illustrated

BLACKIE & SON LIMITED
LONDON AND GLASGOW

BLACKIE & SON LIMITED
66 Chandos Place. London
17 Stanhope Street, Glasgow

BLACKIE & SON (INDIA) LIMITED
103/5 Fort Street, Bombay

BLACKIE & SON (CANADA) LIMITED
Toronto

BOOK
PRODUCTION
WAR ECONOMY
STANDARD

Printed in Great Britain by Blackie & Son, Ltd.. Glasgow

Contents

Illustrations

Useful Hints

1. Join your County Association of Bee-keepers.

2. Attend all honey shows and lectures on bees in your neighbourhood.

3. Take in *The British Bee-keepers' Journal* and as many other bee papers as possible, such as *The Bee World, The Record, The American Bee Journal,* &c.

4. Read all the literature connected with bees and bee-keepers which you can obtain.

5. Make the acquaintance of as many bee-keepers as possible.

6. Order all goods which you will want for the season from dealers in the previous autumn.

7. Remember that, although you will always find bee-keepers ready to help each other, it is everyone's duty to give as well as to receive help.

8. Keep only strong stocks of bees.

BEE-KEEPING FOR BEGINNERS

CHAPTER I

Introductory

Bee-keeping is one of the oldest industries. Before the sugar-cane was introduced into Northern Africa and Southern Europe, honey was commonly used to sweeten food; and in those days a skep or two of bees was to be found in every garden. These straw hives or skeps are still used in some out-of-the-way places; but modern bee-keepers find that they can manage their bees with much greater profit if they use the modern box hive with movable frames.

A stock or colony of bees consists as a rule of one queen bee, which produces the eggs from which the new bees spring; of two or three hundred drones, or male bees, whose use is to fertilize the eggs; and of some twenty or thirty thousand worker bees. These last do all the work of the hive—feed the young; collect from the flowers nectar and pollen, which they mix with the food of the young bees; clean the hive; fetch water with which to dilute the honey for food; feed and clean the queen; protect the hive against robber bees; produce the wax with which they build their wonderful honeycomb; and do many other things necessary to the well-being of the colony.

The Egg (magnified 30 times)

The queen may live for three or four years, but the second year is her best time. If she is allowed to live longer the colony gradually dwindles, unless the bees take matters into their own hands and rear up a new queen for themselves.

The drones live only one short summer. The workers feed them as long as there is likely to be a young queen to be fertilized; but, on the approach of autumn, they are mercilessly put to death, or turned out of the hive to die of cold.

In the height of the season the workers are worn out in from three to six weeks; but those hatched in autumn live longer. Some of them live all through the winter, and start the colony in the spring. Naturally the hard work at that

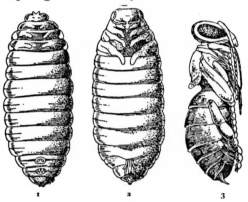

The Development of the Hive Bee

1, The larva full grown; 2, the stage between the larva and chrysalis; 3, the chrysalis. All magnified 5 times.

time very soon wears them out; but if all is well the first born bees of the spring are ready to take their place.

The three kinds of bees all pass through the same changes. First of all the queen lays in a cell a tiny egg, which is fastened by one of its ends to the base of the cell. If it is to produce a drone it is laid in a cell larger than the ordinary worker cells, or, if worker cells are used for drone brood, then the cappings are raised so as to give the drone more room to develop.   But the drones reared in worker cells are not as fine insects as those produced in real drone cells.   Drones, too, are produced from eggs which have not been fertilized, whereas all eggs which are to produce females

Queen   Worker   Drone

The Bee (*Apis mellifica*)

must be fertilized.   The only perfect female in a colony, as has already been said, is the queen. The workers are undeveloped females, and on rare occasions one of these will become an egg layer, but the eggs laid by the " fertile workers ", as they are called, are really unfertilized and only capable of producing drones.   So that a colony which loses its queen and relies on a fertile worker for its increase is doomed very soon to die out altogether.   If the bees wish an egg to become a queen bee a special kind of cell is built round it, which hangs downwards and resembles an elongated thimble. Such an egg, when it hatches into a grub or larva, is fed on a richer diet than either the worker or

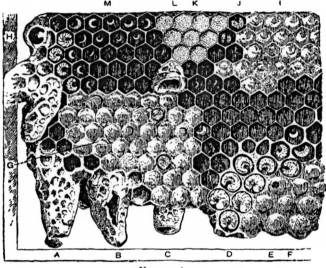

Honeycomb

A, Queen cell, from which queen has hatched, showing lid; B, queen cell torn open; C, queen cell cut down; D, drone grubs; E, drone cell partly sealed; F, drone cells, sealed; G, worker cells, sealed, bees biting their way out; H, old queen cell below which is a sealed queen cell; I, sealed honey; J, fresh pollen masses; K, cells nearly filled with pollen; L, aborted queen cell on face of comb; M, eggs and larvæ in various stages of growth.

drone. But the eggs from which queen and worker spring are of the same kind. We know that this is the case, because worker eggs have been taken and queens have been produced from them by a process of artificial feeding.

The incubation period is three days. Then the egg hatches into a tiny white grub, which lies curled in the bottom of the cell. The workers keep these larvæ warm and feed them, and they soon become erect in the cell, and finally fill it completely. This happens on the eighth day in the case of the queen and the workers, but on the ninth in the case of the drones. The larva, which the workers have ceased feeding, then spins a cocoon round itself and becomes a nymph, which the workers shut up in its cell with a capping of wax. From the cell it emerges a perfect bee. In the case of the queen the whole process takes about fifteen days; for the workers twenty-one; and for the drones twenty-four days.

When the bee first emerges from the cell by biting round the wax capping and pushing it up, the nurse bees near it hasten to help it out, and then feed it and groom it down. The bee is at first sticky, and its body, legs, wings, and feelers, or antennæ, all have to be rubbed and polished. It is amusing in the spring to watch the alighting-board. One bee seems to be attacked by four or five others. One seizes its legs, another its wings, and a third its head.

They all pull and push this way and that, and it looks as if the victim would be torn limb from limb. However, presently the grooms let go, and the bee runs into the hive, clean and ready

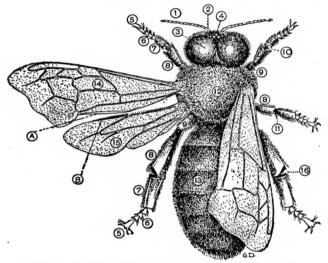

Diagrammatic Enlarged View showing External Anatomy of the Bee

1, Antennæ; 2, basal joint or scape; 3, compound eye; 4, simple eye or ocelli; 5, claws; 6, tarsus; 7, metatarsus; 8, tibia; 9, femur; 10, comb for cleaning antennæ; 11, spine; 12, thorax; 13, abdomen; 14, anterior wing; 15, posterior wing; 16, wax-pincers; B, hooklets which catch on to the fold A in flight and make the two wings into one.

for work. The first work that falls to the lot of a worker bee is to become nurse to the larvæ, and it is several days before they go out to forage.

The worker bees have very efficient stings. These being barbed are often, when used, torn from the body of the bee, which, as a consequence, dies. If the sting has not gone in too far, the bee tries to pull it out again by running round and round so as to wind up the barbs. If it can extricate its sting it flies off a perfect bee, ready and often very willing to repeat the process. The drone has no sting. The queen has a sting which is curved, but she disdains to use it on any but a rival queen.

Stings affect people in various ways, but in all cases it is advisable to remove the sting at once. Do not use pressure, or the poison bag, which is still attached to the sting, will be squeezed and the poison driven into the wound. Place the edge of a penknife under the sting and gently raise it up. It is well to try various remedies until you find out which suits you best. There are a few fortunate beings upon whom bee stings have no effect, but if you do not happen to be one of these, try ammonia, the juice of an onion, or " Milton " gently dabbed on to the part stung.

The drones are large, handsome, burly bees, which make a tremendous noise and move clumsily over the combs. The queen is very

(D 319)

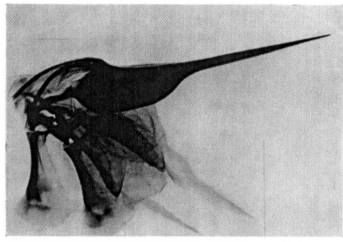

Sting with barbed darts withdrawn from their sheath

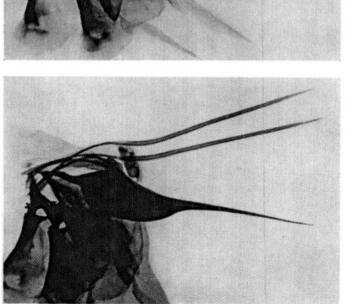

Sting with darts in sheath

Photos. John J. Ward, F.E.S.

THE STING OF THE HONEY BEE (magnified 22 times)

elegant with a long pointed body and short wings; while the ordinary worker bee is so well known by sight that it hardly needs a description, especially in a book intended for bee-keepers.

Is there any other live stock kept by man which satisfactorily manages its own affairs in his absence? Think of all one has to do for cows, sheep, pigs, and poultry. It is sometimes said that bees manage their business best if not interfered with. That is all very well as far as their own business is concerned. They collect, indeed, enough honey to carry them through the winter; but, by careful management on the part of their keeper, they can be made to collect six or seven times as much, and the surplus which they do not need is taken and used by their master.

Bee's Sting (*a*) compared with a needle (*b*), both equally highly magnified.

Consider for a moment the work which bees do for themselves. They ask only for a roof to cover them from the weather. If they are given this, be it hive, straw skep, box, or

even bucket, they are content.   Straightway
they set to work to build their comb, and when
a few cells are ready the queen lays her first
few eggs.   Then, while the masons go on
building, the foragers go out and collect food
from the flowers; the nurse bees feed the
young as soon as the eggs are hatched; the
guards challenge every bee which comes in
at the entrance; and, if the weather is pro-
pitious, very soon there is a stirring colony
of bees.

CHAPTER II

The Swarm

Suppose we possess a hive of bees, strong in number, and that it is the merry month of May. Then, if we have taken no means to prevent it, the bees will swarm.

Somewhere about noon, on a hot, still day, there will arise a mighty buzz, which has a different note from the ordinary buzz of bees, and is called the "swarming note". Out the bees will pour from the hive in a dark-brown stream, flying in wide circles and filling all the air with their colour and sound.

Bees are, as a rule, very good-tempered when they are swarming, and, if you go quietly to the hive and kneel down by the side so as to be out of the bees' line of flight, you may watch the alighting-board and entrance without any fear of stings.

If you are lucky you will presently see the queen come out, for it is the reigning queen which goes off with the swarm. The other bees fly out and begin their circling in the air, but the queen as a rule shows some hesitation. She may rest for a few moments on the alighting-board, and slowly crawl backwards and forwards, so that you can take a good look at her. You will be able to recognize her from the picture and from what has been said about her in the last chapter. Probably she will have a few attendants anxiously following her, and you can compare them with her.

Then quite suddenly she will rise into the air, and the swarming note will be redoubled.

You may watch the cloud of flying bees pass from one part of the orchard or field to another, and presently it will be seen to grow thicker at some special spot. Then you may guess that the queen has settled on a leaf or twig, and quickly the bees surround her.

At first, perhaps, they spread all along a branch; but very soon they cluster and hang in a bunch which has the queen safely in the centre. The size of the bunch depends on the strength of the swarm: a fair-sized swarm would weigh about four or five pounds.

As soon as the bees have settled they should be taken, or they may fly away again. It is best to have a box or straw skep and to hold

A SWARM OF BEES

Photo. J. Blackie

it upside down under the cluster. Then, with the other hand, take hold of the branch on which they have settled and give it a sharp shake or two. Most probably the whole swarm will fall into the skep, which may then be turned up and placed on a cloth, one side being propped up so that the bees may be able to fly in and out. Until the evening the skep should stand near the place where the bees clustered, and then, when all the bees have collected, it may be carried off to the hive. It is well to let the skep stand in the shade, if possible, as bees produce a very great deal of heat when swarming.

The swarm may settle in a very awkward place, such as on a post or tree-trunk or even on a tennis-net. In cases like these it would be impossible to shake the bees into the skep, and they may have to be scooped up in the hands or in a small box and thrown down in front of the skep, when they will probably run in. If once the queen bee is caught and put into the skep, the bees very soon join her of their own accord. If the queen is missed and remains where the bees originally clustered, it is useless to hope to get them to settle; they will return and join her on the bough.

The swarming of bees has always been such an extraordinary phenomenon that, as far back as we can trace the history of bee-keeping, there have been special laws relating to it. Legally, if a swarm of bees leaves ι hive, it belongs to the owner of the hive as l ng as he *keeps it in sight*. Hence he must follow it over hedge and ditch, and even through private grounds, and in fact wherever it goes, or he may lose the ownership of it. No one may prosecute for trespass even if the follower of a swarm enters a private house.

No doubt this gave rise to the old custom of "tanging" the bees. As soon as a swarm left the hive, the good housewife would seize a metal tray or the frying-pan and follow the swarm, banging on her tray with the door key. Neighbours were warned in this way that a swarm was abroad and that the owners knew of it and were following.

Many people begin bee-keeping by buying a swarm of bees in the spring. It is easy to shake the cluster straight into a skep or travelling box, which can at once be fastened up and dispatched by rail or otherwise.

In olden days swarms were welcomed, but now one of the chief arts of bee-keeping is to have the greatest possible number of bees in the least possible number of hives. Consequently the great desire of modern bee-keepers is to prevent swarming. In some years this is possible, but there are years in which the honey flow is poor and when nothing that has yet been thought of can prevent swarms.

Among the reasons why bees swarm are the following: (1) because the queen has not room enough for her eggs; this happens in a good year when many cells are filled with honey; (2) because there is too much honey coming in for the hive to hold; (3) because the queen is old and the bees are anxious to get rid of her and rear another queen for themselves. These reasons at once suggest ways by which we may attempt to check swarming: (1) by giving plenty of room in the brood chamber for the eggs of the queen, and in the supers for the honey; (2) by keeping only young queens in the hives; (3) by cutting away queen cells as fast as the bees build them. If this is done about every ten days, the bees will be content with the old queen, since they have no means of rearing a fresh one.

Swarming means loss of honey, for the bees, which intend to follow the old queen when she swarms, fill themselves up with honey from the old hive before they leave. They are said to take with them enough to last for three days, but swarms have been known to hang for a longer time than that without dying of starvation. This parting feast naturally makes a considerable demand on the stores of the parent hive.

If a hive of bees does swarm in spite of everything, one may either cut out all queen cells and return the swarm to the hive, or, and possibly this is the better plan if the swarm should occur in the middle of a honey flow, one may move the parent hive to a fresh stand, placing a new hive on the old stand, fitted with frames of drawn-out comb, or failing that, with frames of foundation. Hive the swarm into this, giving it a frame or two of brood (without queen cells) from the old hive, and also giving to it all the supers which were on the old hive. The parent hive will hatch a new queen for itself, and will have plenty of brood hatching out too, while the swarm will have all the field bees which will return to their old stand. The swarm will give the surplus for the current year, but the old stock will build up nicely for the following year.

CHAPTER III

Hiving the Swarm

It is best to run the swarm into the prepared hive in the evening, as at that time the bees are naturally looking for a home for the night, and will settle more readily. The hive will have been prepared during the day.

The lower part of the hive or brood box is really a box without either top or bottom, and having across each end a ledge on which rest the ends of the frames, so that the combs themselves hang freely in the brood box. This box stands on the floor board which projects in front, forming the alighting-board. The floor board is generally cut away slightly, so as to form an entrance for the bees into the hive.

If comb which has been already drawn out by other bees can be given to the swarm, it is an advantage; and, if one or more of these combs contain brood, it will prevent the swarm from leaving the hive again as it sometimes does. Failing this, frames fitted with wax foundation must be provided.   These frames can be bought ready fitted, or may be home made.

The wax foundation is a thin layer of bees-wax, which has been stamped with the impression of the base of the cells.   The bees draw this out into the cells, and so save themselves much labour in collecting nectar to produce enough

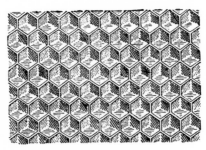

Portion of Foundation

heat to form wax.   This foundation should be wired, that is to say that thin wire should be stretched across the frame and the wax pressed into it, so that when the frame is heavy with honey it will not break down under its load.

Furnish your hive with ten standard frames fitted up in one of these two ways.   Have ready also a calico quilt or sheet with a hole in the middle through which the bees may be fed; also two or three thicknesses of some warm woollen material with which everything may be snugly covered down.

Towards evening carry your skep of bees quietly to the new hive. Place the roof in front of the entrance, resting on the alighting-board: this forms a platform for the bees to rest upon. Remove the quilts and shake about half the bees out of the skep on to the top of the frames. Quickly put the cotton quilt over these to keep them down, and then shake the rest of the bees on to the roof which is in front of the hive. The bees will at once hear the buzz of those already in the hive, and will begin to run in at the entrance.

As soon as the bees have gone down between the combs the cotton sheet can be laid level, and a feeder, containing some thin warm syrup, may be placed over the feed hole. Then cover all over warmly with the woollen blankets; shake the bees which still remain on the roof off on to the alighting-board, and put the roof into its proper place.

So far you will have managed your bees with very little fear of stings, because bees are, as a rule, surprisingly good-tempered when swarming, but in further manipulations it is well for the beginner to have some protection.

A veil is a necessary part of a bee-dress.
This should be of black material; what drapers
call "leno" is satisfactory. It is fairly stiff and
so does not blow against one's face, allowing
bees to sting through; it is also very durable.
The veil should be wide enough to fit easily
over a hat with a fair-sized brim: it should be
hemmed at the top and have an elastic run
through the hem. The length should be enough
to allow of tucking well under the coat collar,
and it should be sewn down the back. It will
be obvious that the bee-keeper's head will be
in a kind of bag, through which he can see,
but inside which he is safe from stings, at any
rate on his face. Beginners sometimes like to
work in gloves, and those made of thick white
leather, with gauntlets to prevent bees from
crawling up the sleeves, are on the whole the
best. For the rest of the costume, one may
say that bees sting dark materials more than
light or white ones, and rough materials more
than smooth, so that a white linen coat or
overall answers perfectly. Sometimes bees
drop off the frames into grass and crawl up
the operator's legs, but boots in the case of
ladies and bicycle clips for gentlemen's trousers
prevent any harm from this cause.

EXAMINING A HIVE

When the bees have adopted their new home a great deal can be learnt of their doings by watching the alighting-board. As soon as the sun shines there is a continual stream of bees passing in and out of the entrance, and the guards can be seen challenging every bee that wishes to pass in. Some of the foragers return apparently bringing nothing with them: these are either bearing water or nectar. Others have large coloured lumps on their hind legs: this is pollen gathered from the flowers and used to mix with the food of the young bees. Almost every flower bears pollen of a special colour, and it is interesting to find out upon what flowers the bees are working by noticing the colour of the pollen being carried into the hives. Early in the spring, when the bees are busy working on the crocuses, we may help them considerably in their task by supplying artificial pollen. Ordinary pea flour answers this purpose very well. We must first get the bees accustomed to visiting a certain spot by putting an old section or frame with perhaps a little honey or syrup in it, then sprinkle a little pea flour on the comb, and very soon bees will be tumbling over each other to get a share of this plentiful supply of the

coveted substance. It is extraordinary with
what rapidity they gather it up, pass it back
under their body, and finally pack it neatly into
the hollows, or pollen bags, on the two hind legs.

Some bee-keepers also supply water to their
hives, but it should not be too close or the

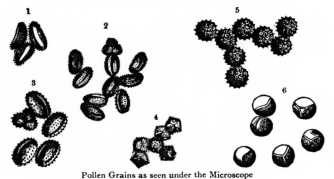

Pollen Grains as seen under the Microscope

1, Water lily; 2, mistletoe; 3, carline thistle; 4, dandelion;
5, horse-thistle; 6, hemp.

bees will fly right over without noticing it.
As a general rule it is not really necessary to
do this. Bees will find their own water quite
easily. I have seen them in a town drinking
from an overflow pipe which dripped slowly.
But if there is reason to suppose that they are
obtaining their water from an impure source,
or if there is any disease in the neighbourhood,
it is well to control the water supply. A large

shallow pan with stones or bits of wood in the water for the bees to alight upon, or better still a covered barrel, which is allowed to drip slowly, can be placed some little way from the hives, and if it seems necessary some good disinfectant can be added to the water.

Plenty of room should be left all round the hive, so that, in manipulating, the operator shall not be cramped for room to place the parts of the hive as he takes them off. If the apiary consists of several hives, they should be at least six feet apart in all directions; thus disease can perhaps be checked before it spreads. The hives should have some shelter from wind and sun, and be well fenced in, so that cattle cannot disturb them.

After it is hived, a swarm should be left severely alone for at least a week before it is examined, as otherwise the bees may forsake the hive, or even kill their queen. They seem to consider that the queen is responsible for any ills that befall them, and when they are in an unsettled state, they punish her accordingly. They never kill a queen by stinging her, but a number of bees collect round her, forming a ball about as big as a walnut, and hug her to death. This is known as " balling " the queen.

CHAPTER IV

Manipulating a Hive of Bees

When a swarm has been in its new home for a week, and has been gently fed during that time, it is safe to examine it, and see what progress has been made. It is often surprising what a strong swarm will accomplish in that short time. Choose a warm, sunny day about noon, and do everything very slowly and quietly, making as little disturbance as possible.

First of all have the smoker going well, giving off a good volume of smoke. If well lighted, the fuel will burn for a long time if the smoker is kept upright, but if it is laid on its side it very soon goes out. Take a loose roll of corrugated paper about the right size for the smoker, and set fire to it at the lower end. Put this into the smoker, and give a few puffs so as to be certain that it is well alight, and then give two or three gentle puffs in at the entrance of the hive. Slowly remove the roof lifts and woollen quilts. Remove the feeder

DRIVING BEES FROM A SKEP

and place it on the ground, with one edge
raised so that the bees are not imprisoned.
Give a few puffs of smoke under each corner
of the calico quilt, and then turn it back at one
side, exposing the division board and a couple
of frames.   Lift out the division board and
place it by the side of the hive: this will give
you room to lift out each frame in turn without

Smoker

hurting the bees.   Loosen each frame before
you lift it, so as not to jerk the bees, and never
give the wax a chance of breaking down by holding a frame horizontal; always keep it vertical.

On one of the centre frames you will probably
find the queen and the beginning of a brood
nest.   There will be eggs in some cells, and in
others the eggs will have hatched into tiny
white grubs (larvæ) which will be curled round
in the bottom of the cell.   As these grow they

straighten themselves out as before explained. Even after a week only, if the weather is good and the flowers are yielding nectar, you will find pollen of various colours in the cells, and also some unsealed honey.

Always replace the frames in the same order, or bees may be crushed by the inequalities of the combs.

If good progress has been made by the swarm it is quite likely that they will soon be needing more room. In that case, when you have arranged the brood chamber, place over it a queen excluder and then either a rack of sections or a rack of shallow frames, putting the quilts on the top of this.

A queen excluder — which, as the name suggests, is to keep the queen from going into the supers and so possibly spoiling the honey by laying eggs in these cells—is made of either zinc, or, better, of wire. The slots allow a worker bee to pass through, but the queen, being larger, is obliged to stay below and keep to the brood chamber. The slots of the excluder should go across at right angles to the frames, and, although the shallow frames go the same way as the standard frames, the sections should go at right angles to them.

When bees have been in a hive for some little time they seal up all the cracks and fasten down the calico sheet with a sticky substance called "propolis", which they collect from the buds of trees.   Everyone knows the buds of the horse-chestnut, with their resinous covering melting and shining in the spring sunshine.

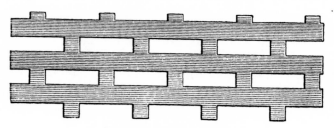

Portion of Queen-excluder Zinc

There is no propolis to collect in the autumn, so that it is a great mistake to open a hive late in the year, as this leaves it draughty all the winter.

We can now imagine that the swarm which we have taken and hived has become a vigorous stock of bees.

Let us turn back for a moment to the old stock from which it came.

The queen-mother, who was probably the parent of all the worker bees in the hive, has flown away with the swarm.   In normal circumstances it is only the queen who has the power of laying eggs, so that, if a hive is left queenless, gradually the brood hatches out and there is no more maturing, for there are no eggs, so that when the present workers are worn out the whole colony must become extinct. This, however, does not usually happen in any ordinary case of swarming. The bees, with their wonderful foresight, have probably built queen cells round some of the eggs and have fed them on " royal jelly ", and in consequence the egg hatches eventually into a queen.

Nature is ever prodigal in her preparations, and the worker bees follow her lead. Not content with having one queen ready to hatch at the time of swarming, they have several, and have even been known to have as many as seventeen to twenty. In due course one of these princesses hatches, and is then known as a " virgin queen ", because she has not yet met a drone and become mated. Certainly she may have been on the same comb as a dozen drones, but mating is only accomplished when flying, and does not take place in the hive.

In a day or two she goes for her mating flight if the weather is fine, otherwise the flight will be delayed. One fine hot morning, after

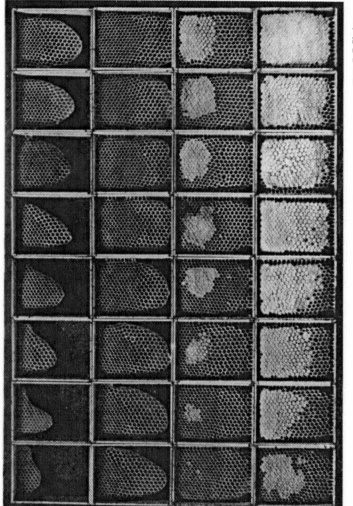

HONEYCOMB SECTIONS IN VARIOUS STAGES

All taken from the same hive at the same time

lingering on the alighting-board and taking short tours round the hive, carefully noting its position and appearance, she darts off into the warm sunlit air.   Every drone in sight darts after her.   Up and up she goes, higher and higher, and is soon lost to sight with her little cloud of drones streaming behind her.   So it is obvious that the prize will fall to the drone which is fleetest on the wing.

If we wait patiently near the hive, she will return, unless some mishap overtakes her. Slowly she comes towards the hive and settles on the alighting-board.   Perhaps we shall see the sign that she is mated, for high up in the air she has been seized by a drone who, in giving her the power of fertilizing her eggs, has also given his very life, for his entrails are torn out, and the queen returns to the hive dragging them behind her.   She will try to get rid of this incubus on the alighting-board before entering the hive, and probably the worker bees will help her.

Presently she enters the hive, a fertile queen ready to be the mother of thousands of bees; for a good queen is capable of laying from two to three thousand eggs a day.   It seems a very monotonous life!   Only twice does she leave the hive in the ordinary course of events, once for mating and later to accompany the swarm. When she was a virgin in the hive, very little attention was paid to her by the workers, but when she comes back mated, the bees hail her as the queen-mother and wait on her in every way, grooming her, stroking her with their feelers, or antennæ, and feeding her by putting their tongues to hers.

This feeding process is easy to see: older bees are very ready to feed young or hungry bees. Each bee puts out its tongue, and the hungry one takes honey from the other. Once mated, too, the queen will probably be allowed to approach the remaining queen cells and sting her rival princesses to death, even before they are hatched.

Sometimes, however, she goes off with part of the bees as a second swarm or cast, leaving one of the later-hatched princesses finally to reign in the hive. Some second swarms have virgin queens with them, and often several virgins are found, but as soon as one of these becomes mated she is allowed to kill off the others.

By the time the first virgin becomes mated much of the old brood will have hatched out, and probably it will all be hatched before the young queen lays her first eggs.   After the eggs are laid three weeks must pass before they hatch, so that the colony often dwindles very much by losing its old foragers and having no young ones to succeed them.   It is at times so depleted that it is all it can do to build up for the winter.

If, however, as soon as a swarm leaves a hive all the queen cells are removed, either the original swarm can be put back or a fertile queen may be bought and introduced, so that no time is lost in producing bees, just when they are most needed for foraging.

The honey flows are of such short duration that every day makes a difference, and unfortunately during an early flow is just the time which a strong colony chooses for swarming.

CHAPTER V

Hives and Appliances

If you send to any of the well-known dealers for a catalogue you may well be bewildered by the long list of appliances described and illustrated in it. Many of these, however, are luxuries and certainly are not necessary for the beginner. Let us first consider the hives. There are many kinds of these on the market, but they all really fall under two heads.

1. The Cottage or Cottager's Hive. This is one which I have already described in Chapter III when we were hiving the swarm. It is single-walled, and the chief point against it is that this makes the brood chamber too cold in severe winters and too hot in good summers. (It is only the brood chamber which is single-walled, since one or more "lifts" or outer covers are used with the supers of sections or shallow frames.) At least one firm of dealers has overcome this objection by making the lift, when inverted, fit down over the brood chamber, so

that in winter it really becomes a double-walled hive. The great point in favour of this hive is its cheapness.

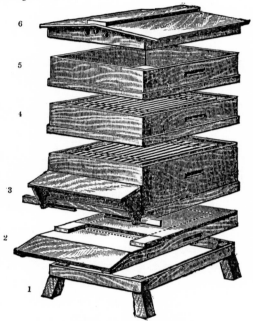

Hive with the Parts separated

1, Stand; 2, floor board; 3, outer case with body-box or brood chamber; 4, surplus chamber with frames for extracting; 5, lift to accommodate section rack when required; 6, roof.

2. The W.B.C. Hive. This is named after a well-known keeper of bees, William Broughton Carr. The whole hive in this case is double-

walled.   The outer case is slightly larger than the inner parts, so that an air space always surrounds the bees.   Since air is a poor conductor of heat, the bees are kept cooler in summer and warmer in winter than in the cottage hives.

In summers of great heat, when cottage hives were in use, cases have been known in which the waxen combs have actually melted down, and the whole contents of the hive, bees, brood, honey, and wax, have all become one sticky mass.   Some bee-keepers fill the air space of the W.B.C. hive in winter with paper or chaff, but this is not necessary and it is really best to let the air circulate.

We have already considered the smoker and also the queen excluder, both of which are necessities.   Some bee-keepers think it best to work without excluders, holding that the ease with which the workers can pass into the supers results in a supply of honey large enough to compensate for what is lost through the brood which a queen may produce in the upper stories.   But anyone who, in a poor season, has had the few sections which would sell at a high price spoilt with brood, will hardly run the risk a second time.

Sometimes a carbolic cloth is preferred to a smoker. This is a cloth soaked in a solution of carbolic acid. As the calico sheet is peeled off the top of the frames, when a hive is being examined, the carbolic cloth is drawn over. The bees dislike this smell so much that it frightens them. The smoke has the same effect: in fact this is the secret of subduing bees so as to handle them with impunity. When they are frightened, their one idea is to protect their precious stores, and they seem to consider that the safest place is inside. Therefore they gorge themselves with honey and then find it more difficult to bend, as they must do if they are to insert their sting.

Hive Tool

Another useful appliance for removing surplus honey is the super clearer. This is a board which fits the top of the brood chamber. It is fitted with a small metal trap called a " Porter" bee escape, which enables the bees to pass downwards from the supers into the brood chamber, but through which they cannot pass up again into the supers.

A tool which becomes indispensable when

once it has been used, is the "hive tool". There are several kinds of these, but the one I mean is formed like a screw-driver at one end, while the other end is widened out and bent under, forming a delightful instrument for scraping the tops of the frames, with very little disturbance to the bees. The screw-driver end is useful for levering the frames apart before removing them from the brood chamber. If this is not done, the bees are oftened annoyed by being jerked.

I have already spoken of the feeders which are used over the feed hole cut in the calico quilt. There are two chief kinds of these: the rapid feeder, which is used for autumn feeding; and the bottle feeder, which can be adjusted so that much or little can be taken at a time by the bees. The latter is used to stimulate the queen to lay eggs, when for some reason the honey flow has ceased, or in the early spring before there is much nectar coming in.

The rapid feeder is a round tin (fitted with a lid) into which the syrup is poured. A hole is cut in the bottom of the tin, and round it is fixed a kind of chimney. The bees from the hive come up through this and drink the syrup. To prevent them from being drowned in it

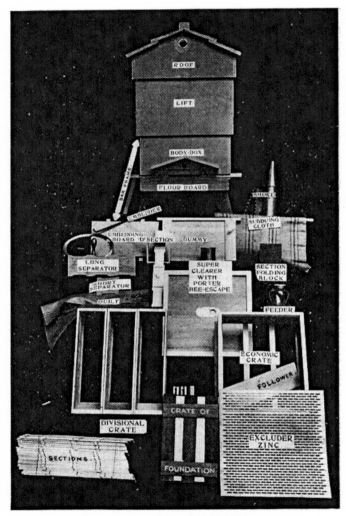

HIVE AND APPLIANCES

a loose cover, which has its top covered with glass, is inverted over it. It is interesting to watch the bees through this glass, while another advantage is that the feeder can be

Plan of Wooden Stand

Bottle Feeder

s is the slot through which the syrup runs; H is one of the twelve perforations in the cover of the bottle.

filled without any bees escaping. It is astonishing how much syrup a strong hive will take down; often two or three pints and more in the twenty-four hours.

It is safest to feed in the evening, when fewer bees are flying, and when there is less risk of robbing.

The bottle feeder is really a glass jar fitted with a tin lid in which ten or more holes are pierced. This is inverted on to a wooden slab in which a tin centre is inserted, and in this tin a slot is cut. If all the holes are over the slot, syrup can be taken from them all, but by twisting the bottle round the holes can be reduced to one or none.

I have spoken of tin as the metal used in the feeders, but aluminium is often used instead; and, although more expensive, it is very much better, for, as the heat of the hive evaporates the water from the syrup, tin is apt to become rusty.

There are other feeders catalogued, e.g. the Canadian feeder, but the two above are those most in use.

A large quill feather will be found very useful for sweeping the bees off combs which have to be removed from the hive for any purpose.

When preparing racks of sections, several points should be attended to. In the first place the wooden part of the sections is bought in the flat, and great care must be used in bending it into shape. If the joints are moistened before bending, it will avoid breakage: either wrap the sections in a damp cloth for some little time before using or dip your finger in water and draw it down the back of each joint. A block can be bought in order to ensure the angles of the sections being true right angles, but this is not strictly necessary if they are folded carefully.

Some bee-keepers give starters only, that is to say, that they use small pieces of wax foundation and let the bees build all the rest of the wax. This is not good policy, as a good deal of the honey, which might go otherwise into the cells, is used in getting up the heat to produce the wax. It is also very difficult to induce the bees to fill the corners of sections unless full sheets of foundation are used. Very thin wax foundation is supplied for sections, since it is eaten with the honey.

CHAPTER VI
General Management

The key-note of good management in an apiary is to have strong stocks of bees at the time of the honey flows.

In rough weather these strong stocks will seize on the few fair hours, and will be so well supplied with willing workers that they will collect enough nectar for the use of the brood and also for surplus; while a weak stock can spare fewer foragers, and often will not secure enough nectar for the ordinary consumption in the hive; indeed it is often necessary to feed very weak stocks even when a honey flow is in progress.

In order to be sure of strong stocks at the right time, it is necessary to know the usual times of flowering of at least the most important plants which yield surplus nectar. Since it takes three weeks for a worker egg to produce a bee, and then a few days before this becomes a forager, one must allow about six weeks in which to build up a strong stock, and this, of course, can only be done at all if the stock is headed by a really good queen, which should be in her second year for the best honey results.

When one starts building up a stock it is wise first of all to remove any excess of stores from the hive for the time being, and give, instead, frames filled with empty drawn-out comb, or failing this frames of foundation. Then use a bottle feeder and feed slowly, using only two or three holes. This creates an artificial honey flow, and the bees, seeing signs of food coming in, feel that they can afford to support a larger population and so urge the queen to lay eggs. This slow feeding is continued until the honey flow is on, and, since by that time the brood chamber will be full of brood and stored syrup, the foragers are obliged to put the booty upstairs, and it only remains for the bee-keeper to see that they have plenty of cells in which to store the honey.

The main flows are from fruit blossom, clover, limes, and heather, and generally speaking these flower in order in May, June, July, and August.

It follows, therefore, that since most apiaries in Britain are not in reach of heather, the surplus for the year is all stored by the end of July or the first week in August. It is well in an ordinary year to remove the surplus honey at this time, and allow the bees to keep, for their winter stores, any more which they may happen to collect. Perhaps one super might be left on, in case of a late flow, but the nectar from late-flowering plants is often poor in flavour, and that from ragwort is positively disagreeable.

In mild seasons the hive should be spring-cleaned early in March or even in February. In order to do this, first lift the brood chamber to one side, placing it corner-wise on a lift by the side of the hive. Then scrape the floor board to free it from bits of wax, dead bees, etc., which have accumulated during the winter. Next scrub the floor board with some good disinfectant; if there is no disease it does not matter very much what disinfectant is used, but Bacterol, Izal, or Flavine are all supposed to be effective when disease is present. Rub the floor board fairly dry, and then lift the brood box back into position. At a later date, when the weather is more settled, a new calico sheet, which has been lately washed in a disinfectant, can be given, and the tops of the frames can be scraped free from propolis, but it is not wise to loosen the quilt from the frames until the weather is really warm.

It is easy at this time to find the state of the stores by the weight of the brood box when it is lifted. Bees and brood weigh very little: it is honey that weighs heavy. If there is reason to suppose that the stores are short, give candy until the weather is warm enough to feed syrup, which in most years would be about the middle of March. Spring feeding should be started as soon as the weather is warm, and a little nectar is coming in from the early flowers, such as willow, celandine, coltsfoot, etc.

It is better to give a little warm, thin syrup constantly than to put on a whole feeder full at one time. If too much is given the bees store it, and possibly the queen may not have enough empty cells in which to lay her eggs. If a little is given at a time it is used by the nurse bees for feeding the brood, and as always, when nectar is coming in fairly regularly, the workers urge the queen to continue her monotonous duty.

Spring syrup should be thinner than that

used in autumn: ten pounds of white loaf sugar to seven pints of water is a good proportion. The making of this spring syrup is not of so much consequence as that of the autumn syrup, since the latter has to be stored in the cells for winter use, while the former is used at once and so there is no fear of fermentation.

Throughout the summer there is little to do except to keep a watch on the stores, look out for swarms or take means to prevent swarming, and give plenty of room in the supers for the bees to store the nectar.

One of the subjects most vital to bee-keepers is this attempt to prevent swarming. As long as the bees swarm just at the beginning of a honey flow, and leave the hive so short of workers that only just sufficient nectar for daily needs can be collected, bee-keeping cannot be profitable. For means of preventing swarms, or rather for attempting to do so (for no certain method has as yet been evolved), larger works than this must be consulted, although the chief methods have been just touched on in Chapter II.

If bees *do* swarm in spite of all attempts to stop them, the best thing to do is to move the old stock to a fresh stand, and either let them hatch out a new queen for themselves, or better still to buy a fertile queen from some reliable source and introduce her to the stock, having first cut away every queen cell.   This cutting away is most important, because, even if *one* queen cell is left, the new queen will either be killed or will go off with some of the bees, forming a tiny swarm.   One can never be quite sure that *every* queen cell has been discovered unless the bees are either shaken or brushed off the combs.

Having removed the old stock, next put a fresh hive in its place and hive the swarm into this.   All the old flying bees on returning from the fields will now join the swarm, the old queen will at once begin to lay eggs, and there will be plenty of room for surplus.   The supers too from the old stock can be given to the swarm with the bees they contain, and a frame of brood with its bees would also add strength to the new colony; but be certain that the frame has no queen cell on it, or the sure result will be another swarm.

If this course is followed it gives the best chance of surplus, as really all that the stock has lost is the number of bees actually looking after the brood on the nine frames which the old stock still possesses.   All the rest, queen, foragers, surplus stores, are in the possession of the late swarm, and their work of honey-getting has not really been much delayed.

Sometimes even the old brood chamber is placed on the top of the uppermost super, after the queen cells have been destroyed.   If this is done the colony is exactly the same as before it swarmed, except that it has more room for the queen in its fresh brood chamber, and the upper brood chamber will give more room for stores, because, as the brood hatches out, the cells are filled with honey, which later on can be extracted.

This plan suggests one of the most usual methods of trying to prevent swarms.   It is known as the Demaree method.   Instead of waiting for the bees to swarm, as soon as ever they are on the outside frames place a fresh brood chamber on the floor board, and in this put the comb which has the queen on it, and next to it, for warmth, another brood comb. Now fill up this chamber with eight new combs; put on an excluder; hunt for queen cells, and put the old brood chamber with two new combs to replace the two removed, either above the excluder or on top of the supers.   When the brood nest has been started in the lower chamber, the excluder can be put above the second chamber so that the queen has the run of twenty brood combs. A good queen will easily keep these filled during the season, and sometimes will be capable of using more.

About the beginning of September begin to feed, if this is necessary, with thick syrup made by dissolving 1 lb. of white loaf sugar in $\frac{1}{2}$ pint of water, bringing it slowly to boiling-point and boiling it for about five minutes. A little salt may be added, and as the syrup cools it can be medicated if desired.

The amount of feeding necessary depends entirely on the weather. In one recent year no feeding at all was needed in the autumn, while in the following one every hive was at starvation point and had about 25 lb. of syrup given to it, as well as candy in the spring,

When autumn feeding is over, the hives should be packed warmly down and left severely alone until it is time to begin stimulating again in the spring; but if, through any oversight, a stock is found to be short of stores, candy, not syrup, must be given during the winter months.

CHAPTER VII

Wintering

Winter is the most difficult time in the management of an apiary, and correct methods are so important that a fresh chapter is justified.

Four things are necessary if a stock of bees is to winter successfuly. (1) A young and prolific queen; (2) a hive *full* of bees; (3) plenty of sealed stores; (4) a weather-proof hive. If these things are assured, unless something unforeseen happens, such as an outbreak of disease, the hive is safe until March.

Supposing, however, that you find yourself, late in the autumn, with one of these factors missing, there are still ways in which things may be remedied.

If the queen needs replacing, it is best to buy and introduce a fresh queen as soon as possible. For it will be too late for the bees to hatch out a fresh queen, and probably the drones will have been already killed off by the workers. It is a sure sign that a hive is not queen-right if drones

are allowed to live after other hives have completed the massacre of their males.

The queen will be sent to you probably by
post in a travelling box, accompanied by about
a dozen worker bees.

This box consists of two compartments, one of
which is filled with candy. The candy not only
nourishes the bees on their journey, but at the
end of the box there is an opening through
which the bees in the hive can eat their way
until they reach the queen and so liberate her.
Full directions for introduction accompany the
queen, and these as a rule are printed on the
cardboard cover of the box, which has to be removed so as to expose the wire gauze which
covers the bees and allows the smell of the hive
to penetrate to them, while the bees are
engaged in eating their way through the
candy.

The old queen should be caught and removed
and the new one introduced with as little com-
motion as possible. Some dealers advise that the
new queen and her attendants should be set free
on a window, and that the queen should then be
caught and then replaced in the travelling box
by herself. I should not recommend a novice
to try this: better to place the cage unopened on

the top of the frames with the gauze side down, and let the bees eat through the candy until they meet.

The directions also tell one not to remove the piece of cardboard which covers the small entrance to the candy box, but I find that the best results are obtained if this cardboard is removed.

After the travelling box has been put into the hive the bees must on no account be disturbed for at least a week, otherwise they are liable to "ball" the new queen. If such a thing should happen it is best to stop operations immediately and close up the hive. The chances are that the queen is then set free, and will be found to be safe at the next examination.

There are many ways of introducing new queens to existing stocks, but they all depend for their success on the concealment of the scent of the new queen. She may be dipped in water and then given to the bees, or rolled in honey taken from one of the frames of the hive. If the queen is not in a cage a piece of cardboard, having some pin-pricks through it, may be placed over the feed hole, and the queen may be kept on the top of this by an inverted tumbler. The bees will feed her through the holes in the cardboard, and, if this is withdrawn the next day, she will be accepted by them.

Should the second factor be lacking, namely, a hive *full* of bees, this can be remedied, even if it be late in the season, by buying "driven" bees. In this case great care should be taken before buying them to ascertain that they are healthy. Driven bees are generally obtained from some cottager who keeps his bees in straw hives or skeps. He wants to get the honey from the skep, and in order to do this he places the skep over a hole in the ground in which he burns a sulphur match.

This "match" is really a piece of brown paper which has had some melted sulphur or brimstone poured over it. It is supported in a cleft stick, which is stuck into the earth at the bottom of the hole, and set on fire. The bees, who have laboured all the summer at their task of nectar gathering, are smothered by the sulphur fumes and fall down into the hole, so the bee-master has his wish—a skep of honey and no bees—but at what a price!

The skeppist is generally quite pleased to let anyone drive his bees for him and, leaving him the skep of honey, to take away the bees with them. In order to "drive" a skep, fix it as

shown in the picture, placing the full skep upside down in a bucket with the empty skep on
the top of it, secured by a couple of skewers or
"driving" irons.   Rap on the sides of the lower
skep with the hands or with two sticks for about
twenty minutes, or until the bees, feeling their
combs tottering, march upwards into the empty
skep.   Every bee can be driven from a skep
in this way, although it may be slow work if the
day is chilly.   When the bees have all collected
in the empty skep, tie a cloth over it and take
the bees home for your trouble.

The chief disadvantages of skeps are that we
cannot examine them to see what progress is
being made; that the combs become old and the
cells filled up with the rejected cases of the grubs,
for, curiously, the bees do not clean these out, and
in time the cells get very small, to the detriment
of the race of bees; manipulation is almost
impossible, and that prevents us from dividing,
uniting, requeening, and from performing other
processes which help greatly in the proper
management of a bee farm; supering for surplus
can only be done to a very limited extent.

One possible advantage of skeps is their
cheapness.   This might induce bee-keepers
who find their bees diseased to burn both bees

and skep, and to make a fresh start and so check the spread of disease. In reality, however, we often find that the old skeps in which bees have died are left standing. In the spring bees rob these out, or swarms from healthy hives take possession of them, and so disease is spread farther afield.

When driven bees have been procured, they may be united to a weak stock in various ways. On the whole the " flour " method seems to be the most satisfactory. By this method you take out each frame in turn from the hive and sprinkle the bees on both sides of the frame with flour, replacing them in the hive. Next, place the roof in front of the alighting-board, as is done in hiving the swarm. Throw the driven bees on to it, and sprinkle them well with flour as they run into the hive. As a rule there is no fighting, and the bees unite amicably. Even if both queens are there they either fight it out between themselves, or it is not uncommon for the bees to keep both queens during the winter, although one of them as a rule disappears in the spring.

Uniting bees is best done towards evening when the driven lot are more inclined to seek a home for the night, and when there is less chance of robbing or interference from other flying bees.

Our third point is that there must be plenty of stores and that these must be sealed. If the autumn has been cold and wet, or if feeding was done too late in the year, the bees may not have been able to cap over their stores. In this case the stores may ferment, and this will most certainly cause dysentery and other digestive troubles. The water from the syrup will also evaporate and cause dampness and mildew in the hive.

The ideal thing to be done in such a case is to remove the unsealed stores, replacing them by frames of sealed stores which have been collected by the bees earlier in the season. Unfortunately, there are not very many bee-keepers who would be so generous as this to their little charges. Most people who find or suspect that their bees are in this condition rely on getting them safely through the winter on candy. It is better to give three or four pounds at a time to each hive, and thus avoid continually peeping to see if more is needed. Every time that the warm quilts are turned back in winter, valuable heat is lost. Every time too that candy is given, the bees become excited and fly, and many are chilled and do not return to the hive, and every bee lost in autumn and winter lessens the chances of having a good stock in the spring. It is a golden rule never to disturb the bees, especially in autumn and winter, unless it is a warm day and they are flying freely.

As for the fourth point, hives should always be weather-proof both in summer and winter. Any defect in this direction can be easily remedied and should have immediate attention. During the winter, the bees are seldom in a true state of hibernation. They are so only in very cold weather. If one has bees in a glass hive, one may watch those on the outside of the cluster gradually working their way towards the centre, and thus the heat of the whole cluster is maintained. The bees nearest to the stores fill their honey sacs, and food is passed gradually from bee to bee until all are satisfied.

Although spring and summer are the busiest times for keepers of bees, yet there is much that can be done in the winter evenings. Everything that will be wanted in the spring can be got ready; frames can be wired and fitted with wax foundation; section racks can be made ready, leaving in one or more "bait" sections from the previous year in order to induce the bees to make an early start; all tin appliances should be scoured and have vaseline rubbed over them to prevent rust; goods should be ordered in good time from the dealers in appliances; and as many books and periodicals as possible should be studied. If one can make hives and appliances much money may be saved.

In such ways the winter months soon pass, and the happy owner is once more among his bees and looking forward to a record honey crop. For all bee-keepers are optimists.

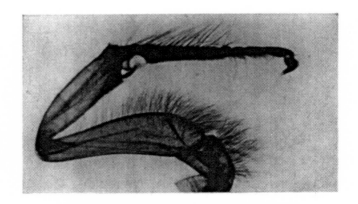

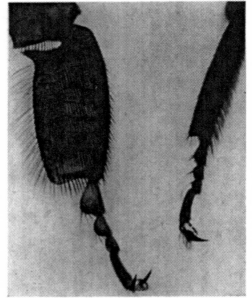

Photos. John J. Ward, F.E.S.

LEGS OF HONEY BEE

At the top is shown the foreleg with the indented joint through which the antennæ are drawn when cleaning them. Below, on the left, is the hind leg of a worker, showing the pollen-basket; on the right, one of its second pair of legs. All equally highly magnified.

CHAPTER VIII

Surplus Honey

To most people the most delightful part of bee-keeping is the taking of the honey! And this may be done with very little disturbance to the bees by using a super clearer. If this is put at night under the super to be removed, the bees will have passed through the Porter escape by morning and the super may be lifted off.

In giving fresh supers it is best to put the new one at the bottom next to the brood chamber, raising the others. The bees fill the top one first, so that this can be removed and a new one at any time given at the bottom.

If comb honey in sections is being worked for, the rack should be removed as soon as the sections are completed, otherwise they become soiled by the bees passing over them; but if shallow frames are used and the honey is to be extracted, the longer it is left on the hive the more mature it becomes and the more delicious it tastes.

It is far more profitable to work for extracted honey, for the same wax can be used by the bees over and over again.

If a rack is removed and the honey extracted, the combs can all be given back, wet with honey, to the bees, who will clean them up and very soon refill them. But in giving wet combs back to the bees do it towards evening, or bees from other hives will be attracted and will probably rob the hive to which the combs are given. When one realizes that the bees use five pounds or more of honey to form one pound of wax, the advantage of using shallow frames is at once apparent.

Honey in sections must be very carefully taken from the rack, and the wooden part of the section scraped clean from propolis. The sections then can be stored away in a tin, and, if kept air-tight, they will last for many months.

In order to extract the shallow frames, first remove the metal ends, as these are liable to get into the extractor. The extractor is a metal cylinder in which two or more frames can be supported vertically. By means of a handle at the top, these can be whirled round so fast that all the honey flies out on to the wall of the extractor. From the walls it drains down

to the bottom, and can be run off by means of
a tap.

After the metal ends have been removed
from the frame, hold the latter over a dish or

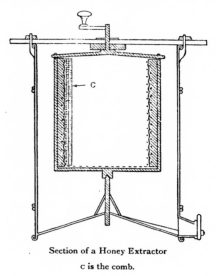

Section of a Honey Extractor

c is the comb.

over the top of the ripener, and, with a hot
knife, cut off the upper part of the cells with
the cappings level with the wood of the frame.
If the knife is really hot and the cut is made
with a sawing action, the cappings come off in
a whole sheet and fall into the dish or ripener
to drain. In order to keep the knives hot have

a jug of boiling water and two knives, which can be used in turn, putting each when done with back into the jug. It is well to wipe the water from the knife each time that it is used, as otherwise the density of the honey is reduced.

When both sides of a shallow frame have been uncapped, it is placed in the extractor. A little practice soon shows at what speed the the honey comes out most easily. When the level of the honey in the extractor is nearing the base of the frames, it should be run off into some receptacle, and then poured into the top of the ripener.

The ripener, like the extractor, is a metal cylinder with a tap at the bottom. It has a funnel-shaped upper part which has a perforated zinc bottom, and it is into this funnel that the shallow frames are uncapped. It is well to tie some muslin underneath the perforated zinc, so as to catch more particles of wax. The honey drips through slowly and collects in the body of the ripener, which holds one or more hundredweights. Gradually the dense honey sinks to the bottom, and so is run off first into the jars. As soon as the honey begins to run at all thin, it should be left in the ripener a few days longer, and, if the ripener stands in the sun or in a warm place, a good density is soon reached by its contents.

If the shallow frames were capped over, the honey will need but little ripening.

A word about honey jars. The usual screw-top jars are not at all satisfactory. They answer fairly well when the honey is liquid, for then the jar can be tipped so that the honey can be reached by a short spoon. One is obliged to use a short spoon, because a longer one would have too large a bowl to go through the neck of the jar. There is a spoon on the market with a prong at the back which hooks on to the side of the jar and so prevents it from slipping in. But why not suit the jars to the spoons instead of the spoons to the jars?

When the honey candies it sets hard in the jars, and a short spoon is of no use. The only way to use the honey is to melt it down: but many people prefer it candied. What is wanted is a shorter jar with a wider neck; something more like a pickle jar. It must be fitted with a screw-cap, because honey absorbs moisture from the atmosphere so readily that the jar would be constantly over-flowing if it were not quite air-tight. There

are some very pretty, fancy honey pots now to
be bought.   Most of them are in the form of a
straw skep.

Another product of the hive which may be
mentioned here is the bees-wax.

It always strikes one as miraculous that bees

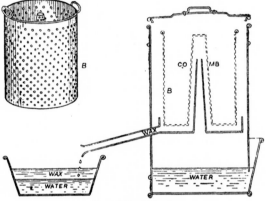

Steam Wax Extractor

B is the perforated comb-basket. The lower pan is placed on a
stove and the steam from the boiling water rises through the
cone and melts the wax.

should be able to produce beautiful white
honeycomb if only they can get nectar.
Huber, the wonderful blind bee-master, kept
bees confined and gave them nectar only, and
the wax was produced.   His experiments, so
thoughtfully planned and admirably carried out,

are well worth following. In the olden days it was thought that the wax was gathered from flowers, but now we know that it forms in scales on the under side of the abdomen of the bee. For wax to appear thus between the segments of the bee's body a great amount of heat is necessary, and, in order to produce this heat, the bees cluster close together and consume a large quantity of food.

As the wax scales form on the clustering bees they are picked off by other bees, and moulded by their jaws and legs into the shape wanted. To obtain what is commercially known as bees-wax, old combs are melted down either by the heat of the sun in a solar wax extractor or by steam.

The solar wax extractor is a useful appliance even if it is not absolutely necessary in an apiary. It consists of a box with a glass lid. This box is lined with tin and has two compartments, separated by a partition of perforated zinc. The box stands slanting towards the sun. All cappings and bits of old comb are put in the upper part of the box, and, as the heat of the sun melts the wax, it runs through the perforated zinc into the lower chamber. A certain amount of honey which was with the cappings will also run through, but since the wax is lighter than the honey, it rises to the top.   At sunset this wax hardens into a solid slab which can be lifted off, scraped and washed,

Solar Extractor

leaving the honey below quite pure and ready for use.

The most common use for bees-wax is to be melted down with turpentine for furniture polish, but it is also used for rubbing on to sewing cotton and for various other purposes.   Formerly church candles were made entirely of bees-wax, but this is no longer the case.

One of the proofs that honey is pure is that after a time, if it is kept in a cold place, it will candy or granulate.

Many people prefer it in this candied state, but, if it is wanted liquid, it can be melted either by standing the tin or jar before a fire or in hot water.   When granulated honey has been melted it does not easily granulate again.

CHAPTER IX

Pollination

In most flowers there are to be found both male and female organs. These are easily seen in regular, widely open flowers, such as the wild rose, the buttercup, or the poppy. The seed vessel of the poppy, or poppy head as it is called, is familiar to everyone. If such a "head" is broken open, it is found to be full of tiny dust-like seeds. This capsule, or head, is the female organ of the poppy, and produces all the seeds.

When the poppy flower opens, one can quite plainly see, surrounding the central capsule, numerous black pin-like stamens. These are the male organs, and the pin-heads contain the pollen or fertilizing dust.

In order that the seeds shall be "set" and capable of producing fresh poppy plants, each seed must combine with one pollen grain. As a matter of fact the term seed should not be used until this combination or fertilization has

The Pollination of the Poppy (see opposite page for description)

taken place. Before this process has taken place, we use the term ovule (little egg) to describe the structure which later becomes the fertile seed.

It is obvious that with the ovules safe inside the capsule, • and the pollen grains in the stamens surrounding it, the chances are against this union being accomplished. In some way the pollen grains must reach the ovules, since the ovules cannot get out of the capsule. In the figure opposite, 1 is the poppy plant with bee approaching the opened flower. 2 is a section of this showing the stamens surrounding the central capsule or ovary, in which are seen ovules on one side and a placenta on the other. 3 shows the ovary and two stamens with a bee crawling across: A, anthers containing the pollen; S, stamens; P, pistil; R, stigmatic ridges on which the bee leaves the pollen from the anthers in its journey across the flower. 4, Ripe fruit capsule cut in half: O, ovary containing the seeds; E, pores by which seeds are shaken out.

Let us see what takes place in the poppy. Everyone has noticed those downy ridges on the top of the capsule which radiate from the centre. Pollen grains are easily brushed off any insect's body by these and held on the ridge.

Notice the bees in the Shirley poppies. They keep up a perpetual hum all through the summer day. There is no nectar in the poppy, but the bees visit it for the sake of the pollen so essential for the food of the larvæ in the hive.

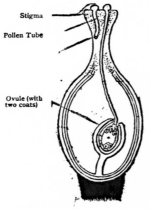

Stigma

Pollen Tube

Ovule (with two coats)

Diagrammatic Vertical Section of Pistil, showing growth of pollen tubes from pollen grains and fertilization of ovule.

The bees busily crawl over the stamens, scraping the pollen from the anthers into their pollen bags. From one side of the flower they go to the other, crossing and re-crossing the capsule, and getting their legs and the under side of their body well brushed by those downy ridges, which in their turn become quite powdered with pollen grains.

Still it is not easy to see how these pollen grains on the ridges can get inside the capsule to reach the ovules. The grain itself does not actually do this, but it puts out a tube which penetrates the wall of the capsule and winds about inside, until it comes into contact with one of the ovules. It pierces and enters the ovule, and then the *contents* of the pollen grain,

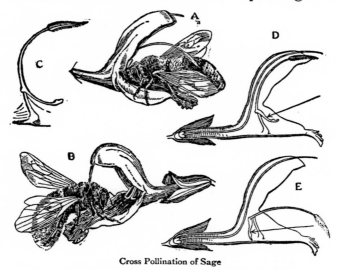

Cross Pollination of Sage

ᴀ, Flower visited by bee, the pollen-covered anther is in the act of striking the insect's back; ʙ, the bee carrying on its back pollen from a younger flower is rubbing it off on to the deflexed stigma; ᴄ, stamens with rocking connective; ᴅ, longitudinal section through flower; ᴇ, same section, the lower arm of the connective lever is pushed backwards and in consequence the pollen-covered anther at the top of the other arm is deflexed. The arrows show the direction in which the bee advances.

which is still up on the downy ridge, pass through this tube and so enter the ovule. The union or fertilization is now complete, and the ovule has become a seed, which after a period of rest will ripen, and be capable of producing fresh poppy plants. When the seeds are ripe, a ring of holes at the top of the capsule open, and, as the wind rocks the capsule from side to side, the seeds are shaken out, and, being very light, are often carried away by the wind to some distance.

We see the bees going from flower to flower seeking for nectar or pollen, but we do not see that they are *buying* it, paying for what they take by performing the invaluable service to the flower of setting its seeds. The flowers seem almost to realize the value of the bee's visit. Watch how the poppy holds up its head and unfurls its four brilliant petals to attract any passing bee. Even the violet, which seems bent upon hiding itself, has such a delicious scent that no bee would pass it by without sipping its nectar.

Most flowers which depend on insect visits for the fertilization of their seeds, seem to have adapted themselves to the shape of the insect. If we keep watch on a sunny summer day by a bed of the common white dead-nettle, we shall not have long to wait before a bee pays a visit to one of the flowers. Even such weeds as the dead-nettle have plenty of nectar to offer.

The bee approaches; the lower lip of the flower makes an admirable alighting stage; quickly she smells the nectar at the base of the flower and in she goes head first with her tongue well out to reach it.   As she plunges in, up goes her abdomen, and a more beautiful fit would be hard to find.   The bee *exactly* fits the flower.   Underneath the hood of the flower are four stamens neatly hidden.   If you bend these back a white thread will spring forward from among them.   In the older flowers this thread is pronged, and it is between these prongs that the pollen grains must get if the ovules are to be fertilized.   Just look for a moment at a young and at an older flower: the stamens of the former occupy the same relative position in the flower as the prong, or "stigma" as it is called, of the latter.   So that when a bee gets a patch of pollen dust on to its back from young flower this patch will be in just the right spot to press against the mature stigma of the older flower and leave some grains upon it. In the case of the dead-nettle the tube which the pollen grains put out has to grow down inside the whole length of the white thread before it reaches the ovules, which lie in their cases right at the base of the flower.

In the willow the stamens are on the tree which is usually called "palm", and their yellow heads light up the whole hedgerow when they are in flower in March and April, and to them the bees flock for pollen which is so invaluable in the spring when the queen is laying eggs at high pressure.   But hunt how we may we shall find no ovules here to be fertilized.

The truth is that this is one of the plants which have their male and female flowers on different trees.   The female willow has short catkins like the male tree, but they are dull and green, and one wonders why the bees should visit them and so bring the pollen.   But, while the male tree offers pollen only to the bees, the female offers nectar.   So the bees fly from one to the other, and the fact that the seeds have really been fertilized is proved later in the year when the female tree opens its seed vessels and sets free its millions of seeds with all their snowy packing of the finest white hairs.

Some plants have the pollen carried by the wind, and in these nothing is produced which would attract insects, and bright coloured petals, nectar, and sweet scent are lacking.   Such plants are the hazel, oak, beech, etc.

There are many curious ways in which

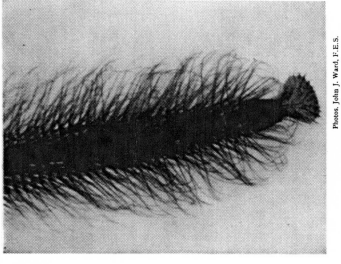

Proboscis and mouth-parts of honey bee
(magnified 22 times)

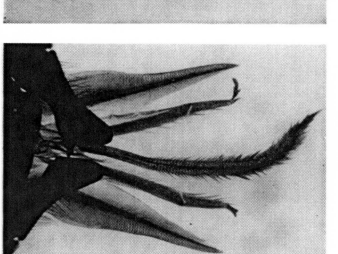

Tip of proboscis still further magnified

Photos, John J. Ward, F.E.S.

MOUTH-PARTS OF THE HONEY BEE

flowers have been modified so that they may, by making use of the insect, fertilize their seeds, but the student must study larger works, such as Avebury's *On British Wild Flowers, Considered in Relation to Insects*, or some of the books by Charles Darwin.

The flowers which yield the nectar of the big honey flows in Britain are first the fruit blossom, and since most of these flowers are a fair size it is not difficult to see for oneself exactly how the bee is conveying the pollen to the ripe stigma. Secondly the clover, but in this case the flowers are too small for one to see what is going on. However, the structure of each separate clover flower which forms the whole head is very similar to the structure of the flowers of gorse, broom, and runner bean, and bees may easily be watched at work on any of these.

The lime, which gives the third flow of the year, is large enough to be investigated, but the curious thing is that, although the lime trees in early July actually hum as if a thousand swarms had possession of them, no seeds are ever set in Britain, and lime trees never spread naturally but always have to be planted.

Heather flowers tell their own tale to the

careful watcher, but the bell heather is the one which is large enough to watch, although the real honey flow comes from the tiny crowded flowers of the later-flowering ling.

The nectar derived from various flowers has many differences, and a good judge at a show will tell you at a glance in most cases from which flowers the various shades of honey have been collected. Heather honey is very dark, and so thick that it cannot easily be extracted, but is actually pressed out of the wax in a honey press.

The honey from common charlock granulates very quickly, and the bee-keeper is fortunate who never gets caught with his ripener full of granulated honey which has to be spooned out or melted down. This candying or granulation is a peculiarity of pure honey and many people prefer it in this state. But it can always be reliquefied if desired by standing the jar or tin in very hot water or over steam.

Fresh clover honey, which is held by many people to be equal to heather honey, looks almost like water, but it granulates to a beautiful smooth white substance.

Sainfoin honey, for which Cambridgeshire is so famous, is a pale lemon yellow, and the wax made from this nectar is also of the same lemon colour.

These are only some of the most obvious differences between various kinds of honey. Every flower bears a pollen grain which can by its special peculiarities be recognized under the microscope; and every flower also yields nectar which has its own distinct taste and smell. The bees appreciate these differences so well that they work one kind of flower at a time, so that an entire pollen load may be of mignonette grains, or from the horse-chestnut, and so on. The nectar, too, is from one source only, if that source is plentiful enough; but if one source runs short, then we get, on extracting, a honey of a medium or darkish colour which is dubbed from "mixed sources".

In most plants a finer seed is produced if the ovule is not pollinated by pollen from the same plant; in fact some apple trees, etc., are self sterile. The cross pollination, so necessary for producing perfect fruits, is brought about by the frequent visits of bees and other insects.

CHAPTER X

Robbing

When the weather is suitable for the flowers to yield nectar, and a honey flow is in progress, the worker bee becomes an insect with one idea only—the gathering of nectar from the flowers. At a time like this one may leave comb wet with honey exposed in an apiary and hardly a bee will notice it.

When, however, the weather changes and the honey flow begins to lessen, the whole character of the bee seems to change. It is as if it says to itself: "Get nectar I must, and if the flowers do not hold it then I must get any sweet syrup from wherever I can." If now we should, by chance, leave a hive uncovered or drop a few pieces of honeycomb near the hives, in a moment myriads of bees have sensed our neglect, and the hive is set upon by robber bees, and the comb we have dropped is soon black with them. Many lives of bees are lost in the mad struggle for honey which follows. If the hive should be a weak one, probably every particle of honey is taken from it; the combs are torn down in the struggle, and many, if not all, of the bees are killed.

There is nothing so devastating to an apiary as a really bad robbing bout. The bees seem to become changed creatures. Instead of our quiet, pleasant, hard-working little charges we find that we have perfect little demons bent on stinging every living creature within reach. Bands of robber bees with a peculiar high-pitched buzz, easily recognized as the "robbing note", go round from hive to hive trying to effect an entrance, and, when once they get a footing in a weak stock, they never leave it while a drop of honey remains.

When the nectar is scarce, bees will enter jam factories, kitchens, and other places where any sweet substances are being used. The honey room where extracting is going on must be most carefully guarded or bees will pour in, and many will be lost, the stocks being weakened just when every bee makes a difference to the chance of the colony wintering safely.

It is obvious, then, that when there is no honey flow in progress, all manipulations should be done towards evening, when fewer bees are flying. Nothing should ever be undertaken which could call the attention of robber bees to a weak hive. All entrances should be closed up to a one or two bee-way, that is to say that only just enough room should be left for either one bee to pass or for two bees to pass one another. Great care should be taken to spill no syrup about, if any hive is having just then to be fed; late swarms sometimes need feeding at such times.

If, in spite of all precautions, robbing should occur, many remedies, which are more or less efficacious, are given in larger books. Probably the best thing is to change the position of the hives, putting the robber hive on to the stand of the robbed one, and vice versa. Sometimes wet grass piled on to the alighting-board and over the entrance may stop the robbers: the bees inside will worry their way through the grass, but the robbers may be daunted by its dampness and cold. It is possible completely to close a robbed hive by day, stopping up the entrance with a wad of twisted grass which will allow sufficient ventilation. In the evening the wad may be removed and the bees allowed to fly, and the hive may be closed again next morning.

Although one may with good fortune be able by some means to stop a robbing bout when it has started, more often a novice finds himself quite helpless. This is certainly one of those cases where "prevention is better than cure", and every means should be employed to prevent robbing from starting.

When hives are moved from one stand to another, as suggested above, it must be borne in mind that bees mark the area in space which the hive occupies and not the hive itself. When the places of two hives, therefore, are interchanged, the bees belonging to the robbed hive will join the stronger robbing colony, and the robbers will carry their booty into the weaker stock and so strengthen it. If a hive is moved from its stand, even for only a few feet, the empty stand will soon be crowded with bees returning from the fields, and, finding no hive where they expected it, they fly aimlessly about until worn out and are finally lost.

If one wishes for any reason to move a hive to a fresh spot, it can be done at any time by moving it about a foot at a time, giving the bees a day or so in between to get accustomed to the new position. This is very slow work. It is better to wait until winter, and then, during a hard frost, when the bees are not flying, the hive may be quietly lifted and moved all the way to its new position. If a little twisted grass is loosely put into the entrance, it will make the bees on emerging realize that some change has taken place, and they will the more carefully mark the fresh spot.

All hives are best placed facing south-east. This allows the sun's rays to pass into the entrance early in the morning; and in many flowers, where the nectar is somewhat exposed, it becomes dried up when the day becomes hot, so that the sooner the bees are roused the more of the precious early morning nectar they secure.

In winter, when one does not wish the bees to be roused up out of their dormant state, hives may face north, or be put behind a wall so that no sun reaches them at all. In America, many bee-keepers carry all their stocks into cellars in the winter, bringing them back to their old stands in the spring; but their winters are very much more severe than ours are in Britain.

If several hives are kept, they should not be closer together than six feet in all directions.

This space will give the operator ample room in which to manipulate, and also, if disease should break out in one colony, it makes it more possible to prevent it from spreading to the others. This distance between the hives also minimizes the danger of the stocks robbing one another. One of the great misfortunes which may follow a robbing bout, is that disease may break out in hives which were previously healthy.

Bees are not the only robbers to be feared. Wasps also prove very troublesome, especially in some years. A weak stock may be entirely robbed out by wasps in a few days. The most satisfactory way of dealing with these pests is to track them to their nests and destroy these with cyanide of potassium. If this plan is not feasible, many of the robbers may be caught by sinking in the ground long-necked bottles containing a sweet syrup. If these are put among the hives they attract wasps, which drown in the syrup; but bees, curiously, do not seem to visit them at all.

Strong stocks can usually hold their own against the invaders, and one often sees many dead robber bees and wasps in front of strong hives, but it is the weak colonies which, as a rule, are singled out for special attack. When the robbers come in force, the bees get discouraged, and seem to let them have their way without trying to resist, and so the colony becomes starved out.

A golden rule for bee-keepers, be they novices or experts, is to keep strong stocks only. It is the weak stocks that cause all the trouble in an apiary. If one has three weak stocks, by far the best plan is to unite them into one strong one; this will then probably give some surplus honey, whereas the three weak ones will most likely give no surplus, and also need a great deal of feeding if they are to winter safely, and even then there is much risk of their being robbed out or dying of cold or disease. Strong stocks are the only satisfactory ones for bee-keepers who hope for success.

CHAPTER XI

Diseases and Enemies of Bees

The subject of bee diseases is a large one, and, at the time of writing, much work is being done by scientists and bee-keepers in order to ascertain the causes of the various diseases at present known to us, and, if possible, to find cures for them.

Dysentery is perhaps the ailment most usually met with, and if the stock be strong this is not as a rule serious. It may be caused by dampness in the hive, produced either by rain leaking through or more often by the presence of unsealed stores. If syrup has been fed to the bees so late in the autumn that they have been unable to cap the cells, then through the winter the warmth of the cluster of bees causes the moisture to evaporate from the syrup, resulting in damp quilts and mouldy combs.

Healthy bees discharge their fæces only when on the wing, and if by reason of bad weather they are confined to the hive for any length of time, the abdomen becomes swollen, and often their very weight makes it difficult for them on first emerging from the hive to fly. We see them crawling about the alighting-board, or crawling up blades of grass in order to gain a point of vantage from which they can more easily take flight. When after some attempts they do fly, their droppings fall on to the hive or the ground near, forming tiny orange spots. After a few frosty days, hives may be covered with such spots, and, if any washing is hanging out near the hives, the garments become stained and need washing again. Those bees which are absolutely unable to get a flight are obliged to discharge their fæces about the hive. After a day or two of fine weather, however, the organs of the bees become again healthy and the trouble passes.

Since prevention is better than cure, be sure that your hives are sound and that the bees are wintered only on capped honey or syrup.

All mould in the hives or on the combs should be most carefully avoided. Many people believe that mould is the cause of the deadly acarine disease of which I must now speak.

For many years the great bugbear of bee-keepers has been what has been known as Isle of Wight disease.   This spread rapidly from hive to hive and whole apiaries were wiped out, causing great loss to expert bee-keepers and preventing many would-be bee-keepers from starting.   Very little was known of its causes or cure, and many conflicting tales were told about its symptoms and results.

It was supposed to be caused by one of the lowest forms of life called *Nosema apis*.

However, Dr. Rennie, of Aberdeen, and others associated with him in his work have now definitely discovered that *Nosema apis* is found in many bees of apparently healthy colonies.   They have also discovered an internal parasite which lives and breeds in the trachea or wind-pipe of the bee, and have named this parasite *Tarsonemus Woodii*, and have named the disease produced acarine disease.   Research work is still in progress, and it is possible that other germs or mites may be discovered; but what bee-keepers at the present time are looking for is a cure. Unfortunately, as long as bees are liable to be wiped out by a disease for which there is no known cure, bee-keeping can never take its proper place as a profitable industry for the villages of Britain. Much can be done by scrupulous cleanliness and a careful use of disinfectants, but the risk of loss is still too great to tempt people to put much capital into a bee business. Bee-keepers are anxiously watching the periodicals dealing with the subject, and are expecting before long to have some definite cure for acarine disease put before them.

If, in the meantime, a novice has reason to suppose that his bees are diseased, his best plan will be to ask the county expert to pay him a visit and give him advice.

The first visible sign of acarine disease is the spectacle of bees crawling about unable to fly. They collect in little clusters for warmth on the alighting-board and on the ground, and, as more and more of them come out of the hive and in their attempt to fly fall off the alighting-board on to the ground, we may get in front of the hive a mass, the size of an ordinary hearth-rug, of crawling bees. This, of course, would be a very acute case.

Other signs are, that the bees become shiny looking, and often run to and fro in a hurried manner or have a trembling motion of their wings and legs. The hive and alighting-board, too, are generally badly stained.

In a bad case of this sort the only treatment is to sweep up and burn the bees, also burning the contents of the hive.   The hive itself may be used again later on if it is thoroughly burned out with a painter's lamp and scrubbed with a disinfectant.   The ground on which the hive has stood should be well dug over and limed.

But a very advanced stage has been reached by the disease before the bees begin to crawl about.   The only way at present to be absolutely certain that a stock of bees is healthy seems to be the frequent microscopic examination of a few bees belonging to it.   This will reveal any traces of *Tarsonemus Woodii.*

A few years ago bees suffered from another serious disease known as foul brood.   Fortunately this is far less frequently met with at the present time.   This disease did not attack the adult bee but the larva, which gradually became a sticky glutinous mass which shrivelled up, causing the capping of the cell to become concave.

In the very early stages of this disease, stocks may be cured by feeding them with syrup medicated with naphthol beta, but in the later stages the only way is to destroy the combs, making an artificial swarm of the bees.

As to other enemies of bees, fortunately they are not many or very serious. Wasps have been dealt with in the chapter on robbing.

Sometimes mice gain access to a hive and make their nests among the quilts, but if all parts of the hive fit tightly and a wire is stretched across the entrance they can easily be kept out. Naturally the wax in the combs, as well as the warmth of the cluster of bees, is an attraction to them.

In the winter, tits are often very destructive to bees. During frosty weather they perch on the alighting-board and tap with their beaks. When a bee leaves the cluster to investigate the cause of the noise, she is immediately snapped up by the tit, which carefully removes the sting before devouring the prey!

One frosty day I counted a dozen bees eaten in this way in about ten minutes. A fine wire netting placed over the whole alighting-board and the porch prevents the bird from perching or tapping, and does not interfere with the bees' flight on fine days.

Another enemy which consumes a certain number of bees is the toad. This animal seats itself on the ground under the alighting-board,

and when a laden bee misses the entrance and
sinks on the ground to regain strength for
another effort it is snapped up by the toad with
almost incredible rapidity.   As a rule there are
not enough toads in an apiary to do very much
damage, and in any case they can be watched
for and removed.

Ants, earwigs, and spiders often make their

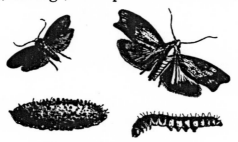

Wax-moths, Larva, and Chrysalis

way into the hives.   Ants may be very troublesome, although the earwigs and spiders do little
harm.   To keep ants at bay, the legs of the
hives may be stood in cups containing paraffin.

A keen look-out should be kept for the wax-
moth. This is sometimes found fluttering about
the hives towards evening.   It tries to find
a way in at the entrance, or conceals itself just
underneath the edge of a lift or the roof.
When a hive is manipulated it quickly flies

among the quilts, and unless it is caught and killed will lay its eggs on the tops of the frames. As soon as these eggs hatch, the grubs work their way from comb to comb, eating the wax and leaving behind them a track of a tough white material. Combs which have been badly infected with wax-moth are completely spoilt for further use.

The novice should keep watch for these pests and take measures against them. Cleanliness and strong stocks of bees are the two best preventives of both disease and enemies.

CHAPTER XII

Treatment of Bees throughout the Year

September. — Just as the gardener's year begins in the autumn, so must that of the bee-keeper.

The surplus honey will, in most cases, have been removed at the end of July. During August the bees should, in average years, collect enough nectar for their daily needs, and in good years they may store some surplus. In heather districts, naturally, a good crop is expected during August, and many bee-men take their hives to the moors for this honey flow.

At the beginning of September it is well to begin a little slow feeding. This induces the queen to lay, and these late-hatched bees are invaluable in helping the colony through the winter. By the middle of the month rapid feeding should be started. The syrup must be made carefully, for it has to be stored for winter food.

Recipe for Autumn Syrup.—1 lb. of loaf sugar to ½ pint of water. Bring it slowly to the boil, and let it boil for ten minutes. Add a pinch of salt, and any other special substance needed at the time, as the syrup is cooling. There may be reason to add a disinfectant to the syrup if there is disease in the near neighbourhood, or if a colony is suspected in any way.

In order to be absolutely sure that a stock has food enough to last the winter, re-fill the feeder every evening as long as the bees will empty it. The stock has ten combs in which to store its winter supply, and when these are filled, having nowhere else to put the syrup, the bees are forced to leave it in the feeder.

If any stocks are weak, they must be united, keeping the best queen. It is little use trying to winter colonies which, in September, cover fewer than six frames. Any queen-less colonies should be united to queen-right ones.

Keep a keen look out for signs of robbing, either by wasps or other bees, and take means at once to stop it. Be very careful not to start robbing by manipulating hives when many bees are flying, but do all necessary operations towards evening. Above all, leave no honey or bits of comb near any hive, for the honey flow is ceasing and a robbing bout may very easily be started at such a time. Reduce entrances to about an inch for the time being.

October.—If any of last month's work has not yet been done, do it without further delay.

Nights are now getting cold, and extra quilts may be given to the weaker stocks.

Remove all superfluous lifts, which can be painted before the spring.

See that all roofs are weather-proof, and guard against these being blown off.

Entrances may still be kept small.

November, December, and January.— All hives should be left severely alone. The owner should know that all is as right as he can make it, and nothing should be done to the bees during these months.

If by any misfortune all is *not* right and there is a risk of losing the colony, then it is best to try and save them. If food is needed, give candy, not syrup. If any manipulation is necessary, choose a sunny day for it, when probably the bees will have broken their winter cluster and may be flying. Entrances of strong colonies may be opened to six inches or even more, but for weak ones this width is excessive.

A hard winter is far better for bees than a mild one. During the latter, bees are constantly flying, which means an undue consumption of stores, and also the loss of many bees which become chilled and do not get back to their hives. Dry cold does no harm to bees, but gives them a good rest before the arduous work of the spring and summer.

February.—With the opening of the lesser celandines and the hazel catkins, bees will be abroad. A warm still day may be chosen to clean the floor board, and candy may be given even if stores are not short, as this will stimulate the queen to lay, and fill the hive with foragers ready to take advantage of the early fruit blossom. Do not disturb the bees more than this. Any manipulation causes excitement and many bees may be lost. These early months of the year, after the inevitable losses of winter, are the most precarious of all, and every single bee is of great value to the future well-being of the stock.

Replace quilts if they become damp, and see that roofs are firmly in position and rain-proof.

If bees have been confined to the hive for

several days in succession, on account of bad weather, do not be alarmed if they stain the alighting-board and front of the hive when they do get a chance of a cleansing flight. A little disinfectant may be sprinkled over the stained parts and round the hive generally.

If not already done, plan out your scheme of bee-keeping for the year, and place orders for appliances and queen bees which you are likely to need later.   On warm days, pea flour may be sprinkled on a piece of old honeycomb and placed among a patch of arabis or aubretia in the sun.

March.—Plum and cherry will be in blossom this month, and since these yield a little early nectar as well as pollen, we can feed a very little warm syrup in the evenings when the weather is favourable. We must, at this critical time, follow very closely in nature's footsteps. If the weather is cold or wet and the bees cannot fly out to forage, then continue the winter treatment—keep them dry and warm, and see that they have candy.   Should, however, the weather be mild and open, so that the bees can fly freely, then we can help them by giving a little warm syrup in the evening, and still continuing the feeding of pollen in the morning. Still continue to disinfect round the hives.

A good rule for March is: "Watch what the bees are doing, and then help them to do it."

April.—This month is often the opening of the bee season, but, on the other hand, it may be only a continuation of winter. In the latter case the winter treatment must be continued, and the bee-keeper must console himself by remembering that a late spring gives more chance of a good amount of surplus from the fruit blossom.

In 1922 there was no warm weather until May, but then began a glorious six weeks with plum, pear, and apple all out together. The surplus from fruit blossom was phenomenal. Unfortunately, however, that was the only honey flow of that year, and bee-keepers, who had not built up strong stocks by feeding in March and April, declare that 1922 was the worst honey year on record. So that, if the weather makes feeding possible, give candy, syrup, and pea flour in April, and build up a hive full of foragers who will repay you a hundred-fold when the apple blossom is out in May.

Still keep the bees warm with plenty of quilts and do not manipulate unduly. On a warm day you can open any hive which you suspect of being queenless, but if a queen or brood is seen, close it up at once or the queen may be "balled". Use very little smoke at this season: it is possible to examine weak lots with no smoke at all. If in any colony the bees are on the outside combs, supers of sections or shallow frames can be put on, or a second brood chamber may be given after the Demaree method.

If any queen-rearing is being undertaken, frames fitted with drone comb should be placed on the outside of the brood nest of the best queen. Drones take twenty-four days to hatch and they should be flying in May to mate with the young queens. It is quite as important for drones to be descended from a good queen as for the virgin queens themselves to have a good mother.

May.—This is the month for surplus from apple blossom. One strong hive will give better results than ten weak ones. The former will store surplus honey, while the latter will use the honey for building up the brood nest.

There are almost as many methods of

managing hives during a honey flow as there are bee-keepers, but one very good way is to place two or three weak lots in their brood chambers one on top of another with queen excluders and supers between. Thus you get a tremendous army of bees on one stand. After four or five days, when the worker bees have marked their new spot, all brood chambers but one can be put back on their old stands, leaving all the supers above the one brood chamber. Be sure that the brood chambers which have been removed have enough bees to cover their brood; if they have not, put some of their brood frames into the hive which has the supers. The weak hives thus still more weakened will probably need feeding.

The bee-keeper during this month has the pleasantest work of the whole year. This chiefly consists in seeing that the bees have plenty of room in which to store their honey. All feeding, except in the case of weakened stocks, may be discontinued.

But let the weather be your guide. Watch the alighting-board closely. You will quickly be able to tell if all is right inside by the general behaviour of the bees. Do not manipulate more than is necessary. Many pounds of honey are lost through upsetting the bees by needless manipulating.

One may expect swarms this month unless means have been taken to prevent them. Remember that if you want honey more than increase it is best to move the parent hive to a new stand, hiving the swarm on the old stand, and giving it all the supers and possibly a frame of brood from the old stock. If increase is wanted, then the swarm may be taken and hived on a fresh stand in the old-fashioned way.

Drones should by this time be flying freely, and nuclei may be made with two or three frames of brood, stores, and bees, together with one or two good queen cells. As soon as the young queen is laying, she may either be sold or the nucleus can be built up into a strong stock for the summer honey flow.

Each section rack should be fitted with one or more "bait" sections, e.g. sections which were drawn out and partly filled by the bees the previous year.

June.—The honey flow from clover is in some places the principal flow for the year, and starts about the middle of June. It often happens that there is a gap of ten days or more between the end of the apple blossom and the beginning of clover. The surplus from fruit blossom may be left on the hives, and the bees can use this, or rather some of it, during the interval; or another course is to remove all supers and feed. Possibly the latter method ensures more rousing colonies for the main summer flow.

The work in June is merely a continuation of that of May. More queens may be reared, more surplus may be extracted, if shallow frames are capped. Sections should be removed as soon as the racks are full, otherwise they are apt to become soiled with the bees passing over them.

Entrances should by now be removed altogether, but keep three or four thicknesses of quilts still on, for the bees will sometimes desert the supers even in June if they are not kept warm.

July.—The honey during this month comes mainly from the limes, and it is unfortunate that, when these trees are flowering, the weather often turns cold and wet. If, however, the weather is favourable, a crop of very good honey is obtained. This is, excepting in the heather districts, the last big honey flow for the year.

Weak stocks must be fed slowly in order

to stimulate the queen, and so produce a large enough population to stand the winter. Surplus may be removed at the end of the month or at the beginning of August, leaving, perhaps, one super partly filled. This, if the weather is bad, will supply the bees with food, and also allow of some more surplus being stored if the weather is good. As the honey flow ceases, great care must be taken not to start robbing, and for this reason it is best to manipulate towards evening, when fewer bees are abroad. Never leave any combs exposed near the hives, and do not call attention to a weak hive by manipulating it more than is necessary.

Narrow the entrances if wasps become troublesome. Kill off any queens which have reached the end of their second year, and either re-queen the stock or unite it to a weak lot having a good young queen.

August.—In this month we hope that the bees will be able to procure enough nectar for their immediate needs and also for winter stores, and if some stores were left to them as directed last month they will be safe. If, however, all surplus is taken, and the weather should turn cold and wet, then stocks must be fed until better conditions prevail.

The nectar collected now is not of such good quality as the early flows, and some bee-keepers keep a rack of shallow frames containing fruit honey, which they give to the bees when the later racks are removed. The bees then take this good honey down into the brood chamber and store it in the empty cells as the brood nest shrinks. For the queen lays far fewer eggs now than she did in May and June. Thus we ensure that our bees have really ripe honey for their winter stores, and we have no need to feed with sugar syrup in September. It is an open question, whether by feeding thus and ensuring good bees, one makes more profit than by selling all the honey and feeding with sugar.

Robbing is still the chief thing to guard against by every means in our power. If wasps are the robbers, they should be tracked to their nests, and in the evening these should be destroyed. Wasps will sometimes make a dead set at some particular hive, giving the bees no chance whatever, and completely clearing out all the stores, besides killing many of the lawful inhabitants.

INDEX

Acarine disease, 96.
Ants, 99.
Autumn feeding, 57.
Avebury, 83.

" Balling ", 33.
Bees-wax, 74.
Beginning, 24.
Bottle feeder, 47, 48. 51.
Brood-box, 27.
Brood chamber, 52.

Canadian feeder, 48.
Candy, 53, 57.
— in March, 105.
Carbolic cloth, 45.
Charlock honey, 84.
Clover honey, 84.
Collecting honey, 67.
Cross pollination of sage, 79.

Darwin, Charles, 83.
Demaree method to prevent
  swarming, 56, 107.
Development of the hive bee,
  13.
Diagram of egg, 12.
Disinfectant, 52, 105.
Driven bees, 61.
Drone, 12, 14, 16, 18.
— comb, 107.
Dysentery, 93.

Earwigs, 99.
Excluder, 56.

External anatomy of bee, 17.
Extractor, 68.

Feeders, 34, 46.
Feeding, 40, 47.
" Flour " method, 63.
Foul brood, 97.
Frames, 36, 65.

Heather honey, 84.
Hibernation, 65.
Hive tool, 45.
Hives, 42, 43.
— in winter, 90.
Honey, 25, 53.
Honeycomb, 15.
Honey flow, 108.
— flows, 51.
— jars, 71.

Isle of Wight disease, 95.

Law, 24.

Manipulation, 34, 62, 104.
— in evening, 88.
Mating, 38.
Mice, 98.
Mildew, 64.
Mould, 94.
Moving hives, 89.

Nymph, 16.